Charles William Penrose

Mormon Doctrine Plain and Simple

Or Leaves from the Tree of Life

Charles William Penrose

Mormon Doctrine Plain and Simple
Or Leaves from the Tree of Life

ISBN/EAN: 9783337499884

Printed in Europe, USA, Canada, Australia, Japan

Cover: Foto ©berggeist007 / pixelio.de

More available books at **www.hansebooks.com**

"MORMON" DOCTRINE,

2

PLAIN AND SIMPLE,
—OR—
Leaves from the Tree of Life.

By CHAS. W. PENROSE.

PUBLISHED BY
THE JUVENILE INSTRUCTOR OFFICE,
Salt Lake City, Utah.
1888.

INTRODUCTION.

There is no subject of popular comment on which there is so little general information as that called "Mormonism." This little work is designed to explain, in a simple way, leading features of "Mormon" doctrine. The terms "Mormon" and "Mormonism" are not strictly correct as usually applied. They are inappropriately derived from the Book of Mormon, which is a work containing the history of the former inhabitants of the American continent, written at different times by various persons and finally compiled by a prophet named Mormon and inscribed upon metallic plates, which were hidden in the earth to come forth in the latter days, for the enlightenment of mankind in relation to the origin of the Indian tribes of this land, and as a testimony that Jesus of Nazareth, who was crucified by the Jews, is indeed the Messiah, the Son of the living God.

This record, giving an account of the dealings of the Almighty with the people it describes, and whose origin and travels, wars and industries, customs and religion, progress and decay it graphically relates, was taken from its hiding place by Joseph Smith in obedience to the revelation and commandment of God and was translated into the English language through a heavenly gift bestowed upon that favored man. Those who believe in the divinity of the book are commonly called "Mormons," and the doctrines which they hold are known as "Mormonism." But it is as inconsistent to call people "Mormons," who believe in the writings of Mormon, as it would be to call them Isaiahs, or Jeremiahs, or Peters or Pauls, because they believe in the scriptures written by those inspired men.

The Church which has been organized under direction from the same heavenly Power that revealed the Book of Mormon, is entitled the Church of Jesus Christ of Latter-day Saints. Its members, then, should not be called "Mormons," but Latter-day Saints. The members of the church established by Jesus and His apostles, as will be seen by reference to the New Testament, were called Saints. The term "Christians" was applied to them in derision, and was first used at Antioch. The members of the restored Church of Christ are called Latter-day Saints to distinguish them from their brethren and sisters of former times. But as "Christians" came to be the common appellation of the former-day saints, so "Mormons" has come to be the title generally bestowed upon the Latter-day Saints, and is used herein only in that sense.

In the twelve leaves which are plucked from the tree of life and herewith sent forth for the healing of the nations from the effects of error and false doctrine, will be found a sweet and sovereign balm for spiritual disorders. And by receiving them, a desire will be created for further gatherings of the same foliage. They will serve to open the eyes of those who have been in spiritual darkness and are yet anxious for the light, and as a preparation for the attainment of that vital power which makes all things new, and quickens and animates earthly beings with celestial life and light.

We ask for the principles here presented, the careful consideration of thoughtful minds, and confidently invite comparison with those utterances of the Jewish prophets and apostles which are contained in the Bible. References to those scriptures will be found at the end of this work, arranged to correspond with each chapter or "leaf."

The young people among the Latter-day Saints will obtain, by a perusal of this little book, an understanding of the fundamental principles of the system which has cost the blood of martyred Prophets and Apostles to establish. And it will be found useful in the missionary field, as a sower of those seeds of truth which, if widely scattered, will surely fall upon some good ground and bring forth fruit for the service of the Master.

We invoke upon this little work the Spirit of the most high God, to whose cause it is dedicated, that wherever it may go light may spring forth to the joy of those who desire the truth, and that by its means many people may be directed into the way which leads to the tree of life, enjoy its luscious fruit and gain the gift of endless lives wherein redeemed man is exalted and the eternal God is glorified.

C. W. P.

CONTENTS.

Necessity of These—The Church Progressive—It Casts Off Evil-Doers—Brotherhood of its Members—Mission of the Church.

SIXTH LEAF.

Apostasy from the Primitive Church—When it Commenced—The Apostles Predicted it—The Apostasy Universal—The Woman Clothed With the Sun, and the Scarlet-Clothed Harlot-What they Represent—The Reformation—Spread of Truth but Lack of Authority—Multiplication of Sects—No Voice from Heaven.

SEVENTH LEAF.

Restoration of the Gospel—Ministration of an Angel—Divine Knowledge and Divine Authority—Keys of Former Dispensations Revealed—Rebuilding of the Church of Christ—The Signs Following—Coming of Elijah—Dispensation of the Fullness of Times—Triumph of the Truth.

EIGHTH LEAF.

Apparent Doom of the Majority of Mankind—No Salvation but by Jesus Christ—Is the State of Man Fixed at Death?—The Common Belief Incorrect—Preaching to the Dead—The Spirit Without the Body Sentient—Nature of Paradise—All People to Hear the Gospel Either in this Life or the Next.

NINTH LEAF.

Decrees of God Fixed in the Spiritual as in the Natural Universe—Ordinances Essential—The Living may be Baptized for the Dead—The Principle of Proxy—The Place for the Administration of Vicarious Ordinances—Revelation of Elijah, the Prophet—Connection With the Spirit World—True Order of Communication—Blessed Results of Work Done for the Dead.

TENTH LEAF.

Universality of Death—Results of the Transgression of Law—Dissolution of the Body not the End of Existence—What is Resurrection?—The Spiritual Body of Jesus—All to be Raised from the Dead—The Order of the Resurrection—Necessity of an Immortal Body—Ignorance of the Laws of Nature—Matter Indestructible—Possibilities of Creative Energy—Life and Immortality Brought to Light.

ELEVENTH LEAF.

Man or Woman Alone Imperfect—Marriage Ordained of God—Sanctity of Proper Sexual Relations—Matrimony a Part of Religion—The First Pair Immortal—Marriage for Eternity—Keys of Celestial Marriage—Condition of Those who Marry Only for Time—Man the Head of the Woman—Plurality of Wives—Continuation of the Righteous Forever—Eternal Family Organizations—Everlasting Increase and Dominion.

TWELFTH LEAF.

Christ's Work Continued After His Death—The Perfect Science of Human Redemption—What was Lost in the Fall—What is to be Regained in the Restoration—Justice Tempered with Mercy—Loss Sustained by the Disobedient—Doom of the Sons of Perdition—The Celestial, Terrestrial and Telestial Glories—Redemption and Glorification of the Earth—Salvation of the Whole Race—The Finished Work of Christ—Universal Dominion of the Father.

"THE LATTER-DAY KINGDOM," a Poem.

APPENDIX.

Scriptural References to all the subjects treated upon in the body of the work.

"MORMON" DOCTRINE,

OR

LEAVES FROM THE TREE OF LIFE.

FIRST LEAF.

Value of Truth—Only One True Religion
—Faith the First Principle—Faith a
Principle of Power—How Faith Comes
—No Man Can Find out God—Deity
Must Manifest Himself—God the Father
of the Race—Personality of God—The
Great Lawgiver Governs Himself by
Law.

There is nothing more valuable than truth. Religious truth, or that which relates to God, our duty to Him, His laws and purposes, and the means by which we may now come to Him and eventually be exalted in His presence, is really priceless. To obtain a knowledge of religious truth, both young and old should be willing to make every exertion and to offer any sacrifice.

There are many systems of religion in the world, but only one can be correct, for the simple reason that there is but one God for the inhabitants of the earth to worship and obey. If there were many true Gods to whom mankind owed reverence there might be several true religions. God is the author or revealer of true religion. Men may invent and arrange methods of worship, imagine and think out doctrines, and formulate and enforce creeds; but they are of no value as a means of salvation. God must be approached and served in the way which He ordains, or the worship and service will not be accepted.

The first principle of true religion is faith. This is the beginning of righteousness. It is the very root of the tree of life, and its sap runs through all the branches. "Without faith it is impossible to please God." And "Whosoever cometh to God must believe that He is." Faith, in its simplest sense, is the assent of the mind, and its assurance of the existence of things unseen by the natural eye. This is belief. In another sense, faith is a motive power, a principle of action. Examination into the secret springs that prompt us in the common affairs of life will show that faith moves us to exertion and incites us to perseverance. It is the assurance we feel of the existence or attainment of things unperceived by the senses, which urges us onward and inspires us with energy. In a higher sense, faith is a spiritual force. It reaches up to the heavenly spheres. It lays hold upon eternal things. It acts upon the grosser elements, and moves spiritual essences and immortal intelligences.

It is in its fullness all powerful. By its exercise God made the worlds, bringing order out of chaos, light out of darkness and visible things out of the invisible, all moved by that spiritual energy called faith. By its power Christ stilled the winds and walked upon the waves, healed the sick and raised the dead. Elijah, by faith closed the heavens, that they rained not, and overcame the might of death, passing with his body into the mansions on high. By faith, Job beheld the coming of the Redeemer, and Paul ascended to the third heaven. And by faith men and women can overcome the influences of earth and time, and rise to communion with angelic beings, and even with God, the highest and holiest of all.

Man must have faith in God in order to become exalted into His presence. No man knows of himself how to reach that position, nor how to obtain salvation from sin and its effects, among which are sorrow and pain, and death as the ultimate. To learn anything in relation to these important matters he must be taught of God, and faith is therefore absolutely necessary in the outset of any attempt to learn of Him.

This faith "comes by hearing," or in other words is developed by testimony. Through the testimony of men divinely appointed to speak in the name of the Lord, faith is awakened in the human heart. It is a principle existing in every soul, but in the condition of fallen humanity is measurably dormant, until quickened by a divine influence. The word spoken by inspired men, accompanied by the influence of the spirit of truth, arouses faith in the soul of man, and by its force he is led to call upon the Lord, and by its light to see his way to repentance and obedience.

No man by his own researches can find out God. He may, by reason and reflection, by observing and pondering upon the wonders of creation, by studying his own internal and external nature, come to the sure conclusion that there is a God, and to a very small extent make an estimate of His character. But without the Almighty manifests Himself in some manner, finite man can never obtain a knowledge of infinite Deity. The speculations of human beings concerning God are many and various, and a vast number of their conclusions inconsistent and vain. Human learning, no matter how extensive, and human research, no matter how profound, are of necessity inadequate alone to the acquisition of a knowledge of divine things. Hence an unlettered person enlightened direct from God, will know more of Deity than the most erudite collegian who has not received this divine illumination.

Some conception of God is necessary to proper faith in Him. On this account He has, at different periods of the world's history, manifested Himself to chosen persons, whom He has deputed to bear witness of His existence and attributes to others, and declare His will and commandments. The history of some of these manifestations and revelations given in olden times is recorded in the Bible. Those that have been vouchsafed to man in the latter times are embodied in what is popularly known as "Mormonism," but which should be called the Everlasting Gospel, renewed on earth.

By these we learn that God is the Father of the human race. As every seed in nature bears its own kind, it is reasonable to conclude that man bears some semblance to the Being from whom he sprang. And this idea is confirmed by the divine declaration that "God made man in His own image." Our Father in Heaven, is then, a personal Being. He is a Spirit. But He is also enclothed in a tabernacle. In other words, He is an immortal Spirit dwelling in an immortal tabernacle. Every faculty and power to be found in mortal man exists in the fullness of its perfection in the person of Deity. Those glorious qualities which make so wide a distinction between man and the lower animals are undeveloped photographs, or rather, embryotic duplicates of the perfected attributes of the Eternal Father.

Being an individual, God, in His personality, cannot be omnipresent. But by the Holy Spirit, which proceeds from His presence and

13

permeates all things throughout the immensity of space, He can see, and know, and influence all things. Yet the Being who has power over all His creations proceeds by law, and while giving laws to all His creations is Himself governed by law and never violates the eternal principles of truth, justice and mercy. The "laws of nature" are the laws of God, and He is consistent with them and those higher laws which pertain to the spiritual spheres.

The Fatherhood of God is a glorious truth that must at some time be impressed upon every one of our race. It involves the brotherhood of man. It is full of ennobling and elevating suggestions, and prompts those who are impressed with its majesty to deeds worthy of so exalted an origin; leads to humility and obedience, and influences all the sons and daughters of the Eternal Father to mutual help, forbearance, charity and affection, as brothers and sisters of a family, whose destiny is connected with the glory, and dominion, and matchless power of the Almighty framer and governor of the universe.

SECOND LEAF.

True Repentance the Consequence of
Faith—Original Sin and Actual Sin—The
Work of Redemption—Universal
Redemption from Original Sin—
Conditions of Salvation from Actual sin
—Baptism, its Object, Mode and Effect
—A New Creature in Christ Jesus.

Faith in God once quickened in the human heart, conscience is awakened and the mind is self-convicted of sin. Repentance follows as the consequence. This includes sorrow for the past and determination for the future. This first of these without the second is not genuine repentance. It is barren and fruitless, and is therefore unacceptable to God. Resolutions of future rectitude are naturally accompanied by grief for past wrong-doing, but regret may exist without reform, and such is not saving repentance, the virtue of which is in turning from evil and cleaving to good. Tears, self-reproaches, lamentations, self-abasement in language or in gesture do not constitute repentance, no matter how loudly they may be indulged in or how conspicuous they may appear, but it is evidenced by forsaking things one knows to be wrong and practising that which one is satisfied is right. Humility is one of its chief characteristics and this prompts obedience.

As repentance follows faith, so baptism succeeds repentance. For the wish to work righteousness in future implies a desire for forgiveness of past guilt, and baptism is ordained for the remission of sins. This opens the broad questions of sin and redemption and the doctrine of the atonement.

There are two general divisions of sin, viz., original and actual. Original sin is that which was committed by the parents of the race, the consequences of which pass upon all of their posterity. Actual sin is that committed by each individual and for which he is personally responsible. Adam and Eve broke the divine law given to

15

them in the garden, the penalty for which was death, natural and spiritual; the first being the separation of the spirit and the body, and the second, banishment from the presence of God. The taint descended to their offspring. Death is the common lot, and a vail is drawn between man and his Maker. Thus mankind are prone to do evil, and the consequence is that "all have sinned and come short of the glory of God." "The wages of sin is death."

Redemption is rescue from the results of the fall. This can only be achieved by the raising of the race from the dead and restoring them to the presence of God. To effect this, Christ came. Doing no sin, He gave Himself as a ransom for those who sinned. He upon whom death had no claim gave Himself to death that he might satisfy eternal justice and give mercy room to act. Death came by Adam, life comes by Christ. Through one act death entered the world, through one act life will come to all that death has grasped. "As in Adam all die, even so in Christ shall all be made alive." Good and bad, believer and unbeliever, male and female, young and old will be raised from the dead and brought into the presence of the Eternal Father. This is the work of Jesus of Nazareth, who shed His blood in this great atonement to redeem all mankind from the fall.

But this was only part of His work. He died not only to atone for original sin but for actual sin, and to become the mediator between God and man. "Without the shedding of blood there is no remission of sin;" this is the law. His blood was shed for the sins of the whole world. For original sin unconditionally, for actual sin conditionally. Mankind had no part in the commission of the original sin, they perform nothing in the redemption therefrom. Its effects came through no acts of theirs; those effects will be removed without anything they may do. No conditions are required as preliminaries to redemption from original sin; it was committed by Adam, it was atoned for by Jesus Christ. But as each person is guilty of his own sins, so he must comply with the conditions which will entitle him to the full benefits of Christ's atonement for his own sins. Among these conditions are faith, repentance and baptism.

Saving faith must necessarily include the Son as well as the Father in its objects, because salvation comes from the Father through the

Son, and as Christ died for all, there is no other name but His given under heaven whereby man can be saved. Repentance, as we have shown, includes humility, which leads to obedience, and baptism follows, in which is given to the repentant believer that remission of sins, obtained through the shedding of Christ's blood in the place of the blood of the sinner.

Baptism as a part of the gospel is the complete immersion in water of a repentant believer, by a man having authority to act "in the name of the Father and of the Son and of the Holy Ghost." All this is essential to its validity. The candidate must believe and repent. The administrator must have divine authority. The ordinance must be performed correctly. There is but "one baptism," as there is but "one Lord and one faith." Any other kind of baptism is spurious and of no effect.

The believing, repentant sinner, after making covenant with God to forsake evil and keep His commandments, is taken down into the water by the duly authorized and ordained representative of the Lord Jesus, and, being dead to his old sins by repentance, is buried from his old life by immersion in the watery grave; and then, raised up again to newness of life, is "born of the water," and stands on earth a new creature in Christ Jesus. He is clean before God. He is as pure from guilt as a new-born babe. Though his sins were as scarlet, he is now washed whiter than wool, and is prepared for the next step on the straight and narrow path which leads to life eternal. Happy indeed is he. Joy unspeakable fills his heart. Peace indescribable dwells in his bosom. Purity shines in all his nature. He has entered by the door, into the sheep fold, and is one of the flock of Christ. The load of his past misdeeds is rolled from his shoulders and he is free. The liberty of the gospel is his. Henceforth he should be the servant only of the King of Kings, and a soldier of the cross.

But he has a warfare to fight which will require all his strength, resolution and fortitude. For he has come out from the world and the world will hate and persecute him, and malign him, and try to despitefully use him. The flesh of his own being will be in conflict with his spiritual nature now brought into actual life. And Satan, the great adversary of the children of light, with his hosts of emissaries will take special pains to tempt and try to allure him from the path of salvation. But God will be on his side, and if he holds true to his baptismal covenants he will come off more than conqueror over all, and obtain the full and complete benefits of the atonement wrought out by the spotless and merciful Savior, who henceforth is his loved and loving Lord.

THIRD LEAF.

The Holy Ghost, its Nature, Office and
Power—Conferred Through the Laying
on of Hands—Gifts and Fruits of the
Holy Spirit—How Obtained—Their
Object and Design—Effects of Its
Withdrawal—Infinite Riches of Its Full
Inspiration.

The repentant, baptized believer arises from the tomb of water
cleansed from sin and washed pure of iniquity. He is spiritually
resurrected. His old life is gone. He is born again. This is a type of
the bodily resurrection to come. He is now prepared to receive the
Holy Ghost, which "dwelleth not in unclean tabernacles."

This is an endowment from on high. It is the Comforter which fills
the absent place of the personal Christ. It is a manifester of truth. It
bears witness of the Father and the Son. It is the light of eternity. It
reveals things present and past, and unfolds events that are to
come. It is the true scripture-maker. By it the prophets wrote the
word of the Lord. It proceeds from the presence of God. It is the
communicating element between man and his Maker. It is the
source from which flow the spiritual gifts of the gospel. Without it
no one can say from knowledge that Jesus is the Lord. Without it,
the things pertaining to immortal spheres cannot be comprehended
by mortals. Without it, no one can see the way which leads to
eternal life. Without it, none can enter the kingdom of God.

There is a set mode by which this great gift is conferred upon
mankind. God's house is a house of order. His laws are set in the
spiritual as in the physical universe, and there is no confusion in
any of His works. The ways of men are not His ways, and He does
not bend them to suit men's diversified notions. To obtain the gift
of the Holy Ghost, the necessary conditions must be complied
with. These we have already explained. The method by which it is
conferred is, the laying on of hands by men who have themselves

received it and have been called of God and ordained to administer it.

True faith, genuine repentance, correct baptism, properly administered, are as surely to be followed by the outpouring of the Holy Ghost, through the laying on of hands, authoritatively administered, as the harvest is to come from seed sown in good soil and ripened by the rains and sunshine of heaven, or as the results of a chemical experiment are to be achieved when the needful elements are correctly compounded.

The effects of this gift upon the recipient are not generally of a startling character. They are not necessarily physical. The chief office of the Holy Ghost is to enlighten the internal man or woman. It administers to the spirit. It brings peace, comfort and joy to the soul. It gives assurance of divine acceptance; and it establishes inward strength to resist sin and evil and lay hold upon all that is good. But it does not convulse the system. It produces no contortions of the countenance. It will not throw people to the earth as if they were dead. Neither will it cause them to yell, shout, jump around in paroxysms or act in an unseemly manner.

Its internal fruits are faith, knowledge, wisdom, joy, peace, patience, temperance, long suffering, brotherly kindness and charity. Its external gifts are manifested in prophecies, visions, discernments, healings, miracles, power over evil spirits, speaking in various tongues, interpretation of tongues, etc.

These several gifts are distributed according to the will of God among the various recipients of the Holy Ghost. One person may receive several of them. Some may not obtain any of those gifts which are manifested outwardly. Neither are the latter always the most to be desired. "Prophecy" is better than "tongues" as a gift, though the latter is more showy, and wisdom and faith are better than either. Divine knowledge with divine wisdom in its use is a gift of priceless worth, bringing joy beyond expression to its possessor, and conferring untold blessings upon others.

These various gifts of the Spirit are obtainable through the prayer of faith. "Ask and it shall be given you" is the promise to the Saints.

And they are called Saints who have obeyed the laws and ordinances we have explained, and received the gift of the Holy Ghost. But their desires must be pure in order to obtain the blessings for which they ask. These are not given as signs to be consumed on any one's lust. Neither are they bestowed as wonders to create astonishment or feed the love of the marvelous. They are designed for the comfort and confirmation of the faith of the true and obedient believer, and as tokens of the love of an indulgent Father, and they must be used in wisdom, or they will be withdrawn and work injury instead of benefit.

The ceremony of the bestowal of the Holy Ghost is called confirmation. As baptism is the birth of water, so confirmation is the birth or baptism of the Spirit. Both are necessary to entrance into the kingdom of God, whether that is viewed in the light of the Church on earth or the glorious dominion of the Father in heaven. Only they who are led by this Spirit are truly the "sons of God."

As it is bestowed through obedience, so it may be withdrawn through disobedience. The condition of those who lose this gift after having enjoyed it is truly lamentable. The light that was within them becomes darkness, and their last state is worse than their first. Their spiritual tastes become dead or vitiated, light seems to them darkness, and that which was once their greatest delight becomes the object of their deepest aversion. They then become a prey to influences of evil; hatred and malice spring up in their hearts towards the children of light; and the culmination of their career, if persisted in and reclamation does not come, is the shedding of innocent blood, for which there is no forgiveness.

The possessor of the Holy Ghost is infinitely rich; those who receive it and lose it are of all men the poorest. But there are various degrees of its possession. Many who obtain it walk but measurably in its light. But there are a few who live by its whisperings, and approach by its mediumship into close communion with heavenly beings of the highest order. To them its light grows brighter every day. For them are joys, anticipations and glorious hopes that thrill no other bosoms, sweet experiences that earthly pleasures cannot bring, and a spiritual growth towards the stature of Christ Jesus that eternity only will fully unfold to general view.

FOURTH LEAF.

Divine Authority—Without it all Gospel Administrations Vain—It Cannot be Acquired—The Priesthood, its Antiquity, Power and Blessings—The Priesthood of Melchisedek—The Aaronic Priesthood—Priestcraft—The Authority of God Must Come From God— Ordination—Value of the Priesthood.

The ordinances of the gospel, being of divine origin, require divine authority in their administration. Baptism at the hands of one not appointed to attend to it is void. It is therefore without value and without effect. If any unauthorized person were to lay hands upon a baptized believer, even if the correct form of the ordinance were observed, the Holy Ghost would not flow to the subject. No matter how good the intentions of either party might be, the lack of authority would vitiate the whole transaction. No company, firm, society, court or government would acknowledge or become responsible for the acts of any but its duly appointed and properly accredited agents. Why then should the Great King endorse the doings of men who take upon themselves duties not required of them, or bestow, through their unauthorized performance, blessings that belong only to the administrations of His chosen ambassadors?

It is strange that intelligent persons who clearly perceive the necessity of valid authority in human affairs, should imagine that it is not necessary in divine affairs; that while no earthly potentate would be expected to pay the slightest attention to proceedings of any pretended representative of a nation or ruler, the Eternal Monarch of the universe must needs honor the acts of any individual of a devotional cast of mind, who chooses to perform ceremonies and ordinances in His great name.

A man may have such faith in God as to obtain choice blessings, behold visions, receive heavenly gifts, and lay hold upon

23

extraordinary spiritual powers, and yet have no right to administer any ordinance in the name of the Lord. Man cannot acquire this authority; it must be conferred upon him in the appointed way.

In every age when the Almighty has had a church or organized body of true worshipers on earth, He has sent among them men who were authorized by Him to act in His name. Of such were Noah, Melchisedec, Abraham, Moses, Elijah, Peter, James and John, and many others. They were not only endowed with the Holy Ghost, but were also appointed and set apart to administer needful rites in God's stead. What they sealed on earth by this authority was sealed in heaven, and what they loosed on earth was loosed in heaven. In other words, what they performed, as directed of God, was accepted by Him and was of the same force as though attended to by Him in person. Any authority less than this is the same as no authority.

This delegated power from God to man is called the Priesthood. Sometimes this term is used in reference to the men who hold this authority. Properly speaking, however, it relates to the office rather than the person. Melchisedec was a great high priest, and the authority he held was eternal in its nature; without beginning of days or end of life. It did not depend upon lineage either of father or mother, and it is written that he who holds it in faithfulness "abideth a priest continually;" that is, he retains it in this world, and also in the world to come. Aaron received a Priesthood which was of another order, and that ran in a family line, descending from father to son, and was subordinate to the higher Priesthood after the order of Melchisedec.

John the Baptist held and administered the Aaronic or lesser Priesthood, but Jesus received and acted in the Melchisedec or higher Priesthood. So John could baptize the repentant for the remission of sins, but could not confer the Holy Ghost as Jesus did. In like manner, Philip, acting in the lesser Priesthood, could baptize the people of Samaria, but had to send for Peter or some other apostle acting in the higher Priesthood, to come down and lay hands upon them, that they might receive the Holy Ghost.

Jesus did not take this authority upon Himself although he was the

Son of God. "He glorified not Himself to be made an High Priest," but His Father called Him, saying, "Thou art a Priest forever, after the order of Melchisedec." Moses and Elijah held similar authority in their day and retained it when they left the sphere of mortality. And they came and administered in that Priesthood to Jesus on the Mount of Transfiguration. As the Father called Him, so called He the apostles, and so, under divine direction, they called and ordained others.

Thus the Priesthood in both orders or branches was continued in the early Christian Church, until through transgression, it was taken from among men, and in its place a spurious priesthood, destitute of divine authority, divine inspiration and divine power, was set up by ambitious and designing men. This is priestcraft, the base counterfeit of the true and heavenly coin.

When the Priesthood is once lost it cannot be regained merely by the hopes, wishes or acts of men. No matter how strong a desire any one may have to benefit his fellow man, he must not attempt to administer to him any ordinance or ceremony of the gospel unless called of God so to do. And this call does not come to men merely "in the heart" or the imagination. A great many enthusiastic persons have felt themselves "called" to the ministry. But this over-anxiety does not give them the Priesthood, any more than strong wishes of a politician for the post of minister to Berlin, clothe him with authority to represent this government in the German empire.

The Priesthood is given by ordination. When there is no man living in the flesh, who holds this authority, its restoration can only be effected by the administration of heavenly beings who formerly held it on the earth. They can return when so permitted and instructed, as Moses and Elias did on the mount. But when the link is restored, they never step over the line of the two spheres for this purpose again, while there remains one man on the earth holding the legitimate authority. For God's house is a house of order, and the rights and powers of His Priesthood cannot be invaded with impunity either by mortal men or the heavenly hosts.

Under divine inspiration and certain rules and provisions, those who hold this Priesthood may ordain others by the laying on of hands.

25

Thus, while mankind are worthy of its administrations and accompanying blessings, it may be perpetuated in the earth, a medium of communication between God and man, a guide for the feet of erring mortals to the straight and narrow path that leadeth unto life. Without it, the inhabitants of the earth wander in spiritual darkness, and those who presume to step forward as their teachers, are blind leaders of the blind, and all their ministrations in the name of Him who never sent them are vain, worthless and without force or virtue in time or eternity.

FIFTH LEAF.

The Church of Christ—Its Unity—
Christ's Church under His Personal
Supervision—Rules of Admission—No
Others Available—Apostleship the Chief
Authority—Other Authorities and
Ministers—Necessity of These—The
Church Progressive—It Casts off Evil-
Doers—Brotherhood of Its Members—
Mission of the Church.

The Church of Christ is an organized body, consisting of those
who believe in Him and have shown their faith by obedience to the
initiatory ordinances of His gospel. It may contain many branches,
but they will all be connected with the main body, and will all have
the same characteristics; that is they will hold the same doctrines
and be animated by the same spirit. No matter how many sections
of the Church there may be, or how widely they may be separated
geographically, they will be governed by the same rule of discipline,
and be under the direction of the same head.

The Church of Christ must be established under His own
supervision, and according to His commandments. A society of
persons professing to believe in Him, but organized without any
communication from Him, is not and cannot be His Church,
whatever its members may call it, or however sincere they may be
in their intentions. Some suppose that every one who believes that
Jesus is the Christ is, by virtue of that faith, a member of His
Church. This is a palpable error. As well might it be assumed that
all who believe that the Order of Masonry is a correct form of
brotherhood, are by that belief made members of the Order.

All societies have some established regulations for the admission of
members. The Church of Christ is no exception to the rule. But the
initiatory rites in His Church are prescribed by Him, and no one has
the right to change them or substitute others in their place. They

27

are uniform for all people, of both sexes, of every race and of every grade of society. The churches established by men have various modes of receiving members and of conducting church government. This is one proof that they *are* the churches of men, and not of Jesus Christ.

We have already explained the first principles of the gospel which must be received and obeyed in order to obtain a standing in Christ's Church. Those who have believed, repented, been baptized by one having authority for the remission of sins, and have been confirmed by the laying on of hands for the gift of the Holy Ghost, are thus made members of the Church of Christ. And this is the only way of admission. All who have not complied with these rules are outside of the Church and can get in by no other door than this appointed entrance. Christ will not accept the devices and ordinances and ceremonies ordained of men. They are not His, and are of no force or effect so far as the kingdom of heaven is concerned, either in this world or in the world to come.

Christ is the head of the Church, as man is the head of the woman. But as the woman has also a head to her own personality, so has the Church. The apostleship is the principal governing authority thereof. When Christ ascended on high, the earthly headship devolved upon His apostles, of whom Peter, James and John were the chief. There were also the seventy appointed by the Savior as His traveling ministers, and He gave other officers to the Church, such as evangelists, pastors, elders, bishops, teachers, deacons, etc. All these were under the direction of the apostles, who were inspired, and instructed, and led by Jesus, even after His ascension, and were filled with the Holy Ghost, which bears record of the Father and the Son.

A church which has not inspired apostles nor prophets, cannot be the Church of Christ, for these are essential to its full constitution. All the officers we have named are necessary, in their various positions, to the complete organization of the "body of Christ."

Through these appointed servants of God, the members of the Church are instructed in their duties, led along in the path of truth, admonished of their faults, rebuked for their transgressions, brought to the unity of the faith, corrected of their errors, and when they become evil-doers, and reformation is not probable, disfellowshipped from communion or excommunicated from the Church.

The Church of Christ is progressive. That is, it advances in the knowledge of the truth. As fast as its members are prepared for additional light, through the practice of principles already revealed, new manifestations are given, for the growth of all who will receive the truths unfolded towards the fullness of the stature of Christ Jesus. Old truths are not discarded, but new truths are added, and clearer light is thrown upon what was previously known. Thus the Church advances and prepares its communicants for a higher sphere when they pass away from the plane of mortal existence.

But while it casts off no truth, it eliminates from itself, by natural process, everything obnoxious to its health and vitality. Corrupt and wicked persons occasionally find their way into its sanctuary; some, after being washed from their impurities, turn again to their filthiness, and others become rebellious and discordant. These incongruous elements are gradually separated from the body. For the Church is a living thing, and casts off that which does not assimilate or which is inimical to its growth, harmony and progress.

The members of the Church are all united by a fraternal bond. They are all brethren and sisters, no matter what their condition in life, no matter of what nationality. Indeed nationality is swallowed up in fraternity. They are no longer Jew or Gentile, English, German, Danish or American, they are all one in Christ Jesus. They are no more Catholics or Protestants, Dissenters or Episcopalians, but are baptized by one spirit into one body, and in all essential principles have one faith, and are joined together in the same mind and the same judgment.

The Church of Christ in this and every other age, is connected with the Church of previous ages. That portion behind the vail works in harmony with the new Church in the flesh, and its members, whether in the body or out of the body, move to the common end: the establishment of the kingdom of heaven upon the earth, the spread of the principles of the true Church, until "every knee shall bow, and every tongue confess that Jesus is the Lord, to the glory of God the Father."

Then the Church will have filled its mission—to preach the gospel, administer in its ordinances, unite the Saints, manifest the things of God, establish righteousness, bring together the heavens and the earth and make straight the path for the Lord Jesus. And the vail of the covering will be taken away; the Church of the Firstborn will be one in all things beneath and above; evil will be swept from the earth; and truth, peace, harmony and praise will glorify this planet and its inhabitants, who will know God, from the least even unto the greatest.

SIXTH LEAF.

Apostasy from the Primitive Church—
When it Commenced—The Apostles
Predicted it—The Apostasy Universal—
The Woman Clothed With the Sun, and
the Scarlet-Clothed Harlot—What They
Represent—The Reformation—Spread
of Truth but Lack of Authority—
Multiplication of Sects—No Voice From
Heaven.

Comparison of the various sects of modern Christendom with the
Church of Christ as established by Jesus and His apostles, which
was briefly described in the preceding chapter, will show that there
has been a wide and remarkable departure from "the faith once
delivered to the saints." It is contrary both to scripture and sound
reason to think that Christ would set up two or more discordant
religious systems to distract mankind and cause strife and
contention. "God is not the author of confusion." There is but one
straight and narrow path that leadeth unto life. The mind of God is
one; the minds of men are various. The fact then that there are
various opposing religions in the world is conclusive evidenced that
men have been engaged in their invention. It is also clear that they
have established very imperfect imitations of the true Church of
Christ.

The departure from the order, doctrine, ordinances and spirit of
primitive Christianity commenced at a very early period.
Contentions began to creep in among the early saints, and they
soon commenced to array themselves in factions, some being of
Paul, others of Apollos, others of Cephas, etc. And the inspired
leaders of the Church foresaw the great apostasy which would take
place, as may be seen from their epistles.

Paul declared that the day of the Lord's second advent would not
dawn until a "falling away" should occur. He described the

31

condition of apostate Christendom, when the people "would not endure sound doctrine," but would "heap to themselves teachers, having itching ears;" when "doctrines of devils" should be taught instead of the pure gospel; when they would have "a form of godliness, but deny the power thereof;" and Peter declared that false teachers would arise in the place of the duly authorized servants of God, and bring in damnable heresies; who "through covetousness would make merchandise" of the souls of men; and by whom "the way of truth would be evil spoken of." This "mystery of iniquity" had already begun to work even in their day, and rapidly increased after their departure.

The combined powers of the world, the flesh and the devil, made such inroads upon the Church of Christ, that, by the time when John, the beloved disciple, was banished to the isle of Patmos, where he received the great vision known as the Book of Revelation, only seven branches of the Church were worthy of divine mention, and some of them had become so corrupt that terrible denunciations were hurled against them, and they were threatened with complete rejection.

In that same vision the inspired apostle beheld the utter and universal apostasy of the Church and the spread of spurious Christianity until *all nations* were "made drunk with the wine of the wrath of the fornication" of Babylon, "the mother of harlots and abominations." Instead of the chaste Church of Christ, clothed with the sun, the moon under her feet and the crown of twelve stars upon her head, the scarlet-clothed impostor, sitting upon the beast, grasping a golden cup full of filthiness for the whole world to drink. Regal pomp and state power, instead of the solar glory of the Melchisedec and the lunar rays of the Aaronic Priesthoods, with the stellar crown of apostleship shining at the head! Mystery instead of light! Painted gaudiness instead of modest purity! Names of blasphemy instead of that one sacred name at which every knee should bow!

When the lights that Christ kindled on earth to lead mankind in the only true way were put out by the hands of murderous men, darkness overspread the world, and "gross darkness covered the people." Errors multiplied. Heresies sprang up like rank weeds. The

Spirit of Christ gradually withdrew. And when what was left of the form of Christianity became allied to the softened paganism of the Romish empire, the angels looked down from afar upon another triumph of the arch adversary, who rules as prince of this world, and reigns in the hearts of the children of disobedience.

The Papal church, seated upon the Romish State, was fitly prefigured by the woman upon the beast. The Church of Christ was gone, without even a shadow of its presence to be seen upon the earth. All nations were blinded and intoxicated by the mystery and abominations, the heresies and perversions, the pomps and vanities of this spurious ecclesiastical system, with its popes and cardinals in the place of apostles and prophets, its priestcraft in the place of the Priesthood, and its force, bloodshed, cruelty and lust in the place of the love, liberty, peace and charity of the departed Church of the Redeemer.

After a time came the reformation. Protestants against the tyranny, falsehood and gross villainies of this blasphemous hierarchy sounded aloud the story of her abominations and shook all Christendom with the force of their eloquence. Anathematized and excommunicated from the mother church, they established new churches, discarding many errors but retaining as many more. Still further "reformations" were inaugurated, originating more churches, and thus sects produced sects, and as religious liberty increased so religious systems multiplied, until the term Christianity covered an incongruous mass of discordant elements, representing all shades of human opinion, without a single authoritative voice deputed of heaven to harmonize and bring them into order.

For, though immense good accrued to the world through the exposure of error and the unfolding of truth, which were the consequence of the reformation and its successive developments, and though many excellent mea spent their lives and suffered cruel deaths for principles of righteousness, yet there was no direct communication established between them and the heavens, and that authority by which the apostles administered for and in behalf of the Father, the Son and the Holy Ghost was still unrestored to man. There was no inspired prophet, no gifted seer, no appointed revelator through whom the will of God could be made known.

Therefore, the ordinances of the gospel could not be administered acceptably to God, and all such ceremonies as were established among the various sects were of necessity void and without virtue in heaven.

So the world rolled on, and men framed religions, all containing some truth as well as some errors, and many persons who would have done well in advocating what they believed to be right, in their own names, undertook to assume the name of the Trinity, and to officiate as though authorized by Jesus Christ, while they openly admitted that there had been no communication from on high for centuries, and maintained that the days of revelation were gone forever.

And thus the effects of Mystery, Babylon, the Mother of abominations, were felt directly or indirectly throughout all the nations professing to be Christian, and millions upon millions of mistaken souls passed behind the vail without receiving the principles and ordinances of salvation, and the living and the dead were left in the spiritual darkness of centuries of apostasy to wait until the dawning of the great and last dispensation, the times of restitution, when the crowning act of God's mercy to man should be performed, and the ushering in of the millennial day should bring again to the world, with increasing glory, the gospel, the Priesthood, the blessings and the powers of all former ages, for the salvation of the human race and the permanent establishment of the Church and kingdom of God, no more to be thrown down forever.

SEVENTH LEAF.

Restoration of the Gospel—Ministration of an Angel—Divine Knowledge and Divine Authority—Keys of Former Dispensations Revealed—Rebuilding of the Church of Christ—The Signs Following—Coming of Elijah—Dispensation of the Fullness of Times—Triumph of the Truth.

Having shown the universal apostasy from the Church established by Christ and His Apostles, we now turn with pleasure from the dark picture of error, strife, confusion and priestcraft, painted in sombre hues during a long succession of centuries, to a more cheering and truly delightful subject.

The same inspired apostles who foretold the general departure from the "way of truth," also predicted the restoration of the gospel, the ushering in of a later and final dispensation, and the ultimate triumph of God's kingdom upon the whole face of the earth. After seeing the dominion of the mother of abominations extending to all the kingdoms of the world, John, the beloved, beheld her entire destruction. This was preceded in the vision by the coming of an angel from heaven with the everlasting gospel for every nation, kindred, tongue and people, and the cry from heaven, "Come out of her, my people, that ye be not partakers of her sins and that ye receive not of her plagues."

We are able to state, with the most positive assurance, that the angel with the gospel has come, and that the voice from heaven has been uttered as a warning to all nations; that the gospel will be preached and the warning will be sounded, by divine authority, to every tribe, and nation, and tongue. Joseph Smith was the chosen instrument in the hands of God to receive the glad message and direct its promulgation to all the world. Angels do not travel and preach to mankind in person; when they bring tidings from on high they deliver the heavenly mandates to a chosen man who, in turn, makes them known to his fellows.

But though the ministry of angels is not general, all people may know thereof of a surety by obedience to the commandments revealed, which is followed by a divine witness of their truth and of the fact of the manifestation. Thus, while Joseph Smith was selected to receive direct divine communications, every one who in faith obeys them, obtains a satisfactory testimony that the message is true and that the messenger was authorized to declare it.

But receiving the gospel, whether by angelic ministrations or otherwise, is one thing, and obtaining authority to preach it and administer its ordinances is another. Knowledge, light and revelation may be enjoyed, and yet the favored recipient of these blessings may be without any authority to perform any official act in the name of the Lord. Joseph Smith not only received the ministrations of the angel bearing the everlasting gospel, but also obtained the right to officiate in all its ordinances, rites, ceremonies and endowments. He did not receive this authority from man. As we have already shown, it had departed from the earth centuries before. No amount of learning would bring it. No college, prelate, potentate or priest could confer it. All the wealth of the world could not purchase it. It does not come by the will of man.

How did Joseph Smith gain it? Holy men of old, who held the keys of this power in former dispensations, came to earth as ministering spirits and ordained him to the same offices which they held in mortality. First came John the Baptist, who was beheaded for the truth's sake, bearing the keys of the Aaronic or lesser Priesthood, and ordained Joseph Smith and Oliver Cowdery to the authority thereof, with the right to preach the gospel of repentance and administer baptism for the remission of sins. But as John did not hold the power when on earth to confer the Holy Ghost, he did not presume to bestow it upon others. Next came Peter, James and John with the keys of the Apostleship, of the holy Melchisedek Priesthood and of the dispensation of the fullness of times, which they conferred upon Joseph and Oliver, giving them authority to ordain others to this ministry and to confirm baptized believers by the laying on of hands for the gift of the Holy Ghost.

The lesser Priesthood holds the power of the ministration of angels and authority to administer in temporal things. The greater Priesthood holds the power of communion with the Highest and of attending to all things, spiritual and temporal; for the salvation and exaltation of man till he reaches the actual presence of the Eternal God, and shines forth in the fullness of the attributes of his Almighty Father.

Thus power was restored to rebuild the Church of Christ, to preach the true gospel; to baptize penitent believers for the remission of sins; to bestow upon them the Holy Ghost, bearing witness of the Father and the Son and of acceptance with them; to appoint and ordain all the various ministers necessary for the publishing of the truth to all nations, the work of the ministry, the perfecting of the Saints and the edifying and government of "the body of Christ."

So the Church was set up in these latter times. Humble believers received the word with gladness, and obeying it, obtained from God the witness of its truth. The signs promised to believers followed them. They spoke in other tongues, prophesied, saw visions, dreamed divine dreams and enjoyed all the gifts of the Church as did the saints of old. The sick were healed by the laying on of hands, devils were cast out, the deaf heard, the eyes of the blind were opened, the lame leaped for joy, the tongue of the dumb was loosed, the heavens were opened to human view, and the Holy Ghost, as on the day of pentecost, rested down in power upon the Saints of the new dispensation.

Then they *knew* for themselves. Doubt had fled, the darkness was dispersed, Satan trembled, priestcraft raged, and while the tidings of the restored gospel caused joy in heaven and praise on earth, the powers of evil in and out of the flesh conspired to fight the truth, make war upon believers and persecute the servants of God unto death. But the Lord strengthened the hands of His people and poured out light and knowledge from on high. The hidden things of ages were brought forth. Revelation after revelation was multiplied to the Church.

Then came Elijah the Prophet, bearing the keys of the turning of the hearts of the fathers to the children and of the children to the fathers, that the link of the broken chain of the Priesthood through the ages might be welded together, and the spirit world be known to men in the flesh. Next came Moses, the man of God with the keys of the gathering of Israel, that the remnants might be brought in from their long dispersion and inherit the lands promised to their forefathers. And Raphael and Gabriel and other holy messengers also appeared, each in their order, bearing the keys of their respective ministries when living as men upon the earth, that all the powers needful for the establishment of the great and last dispensation of the fullness of times might be centered upon the head of the man chosen to open it to the world, and that he might bestow them upon others called and chosen by the spirit of revelation.

Glory to God in the highest! The straight and narrow way is opened. The silence of ages is broken. Jehovah speaks from out the bosom of eternity. Angels again come down from the abodes of bliss. Communication is restored between man and his Maker. The Holy Ghost again comforts, reveals and bears witness. The sacred gifts are once more enjoyed. All earth shall hear the glad tidings. Every soul shall be warned. And though Joseph, the chosen seer, and many of his brethren have become martyrs for the truth's sake, and the bosom of mother earth is stained red with the blood of the persecuted Saints, the Church re-established, the Priesthood restored, the truths now revealed shall never, be taken from earth again, but they shall spread and increase and prevail and triumph, until darkness and evil, and sin and Satan shall give way, and this planet, ransomed and redeemed shall be crowned with the glory and presence of its rightful King, Jesus the anointed, the sinless Son of the omnipotent God.

EIGHTH LEAF.

Apparent Doom of the Majority of
Mankind—No Salvation but by Jesus
Christ—Is the State of Man Fixed at
Death?—The Common Belief Incorrect
—Preaching to the Dead—The Spirit
Without the Body Sentient—Nature of
Paradise—All People to Hear the Gospel
Either in this Life or the Next.

One of the great difficulties in the way of inquiring minds, desirous
of understanding gospel truth, is the apparent doom of the great
bulk of the human family to perdition. The declaration is plainly and
positively made in the scriptures that there is no other name given
under heaven whereby man can be saved, but the name of Christ
Jesus. It is also proclaimed that "except a man be born of water
and of the Spirit, he cannot enter into the kingdom of God."

Many millions of the earth's inhabitants have passed away without
hearing the name of Jesus, or having any opportunity of the
privilege of the second birth. And the query arises, must all these
souls be lost in consequence? And if so can the God of the Bible be
just? Further; the question comes up, If the world has been in error
so long, and the Church of Latter-day Saints is the only true
Church of Christ, what has become of the generations of
professing Christians, who lived and died in the centuries between
the loss of the gospel and the Priesthood and their restoration in the
present age?

The difficulty arises through lack of a correct understanding of the
plan of salvation, and through the erroneous doctrines of
unauthorized teachers. Orthodox "Christianity" affirms that the
future state of man is fixed at death; that the departing spirit goes
either to an eternal heaven or an everlasting hell; and that there is no
possibility of change, but, to use a familiar saying, "as the tree falls,
so it lies." The light of modern revelation rolls back the darkness of

39

ages and unfolds the glorious plan of human redemption in its fullness, and the illuminated soul perceives the triumph of justice in union with mercy, through the extension of gospel privileges beyond the narrow sphere of this mortal life.

Why should the opportunity to learn and the power to obey the truths of the gospel be confined to dwellers in the flesh? Is it to be supposed that when the immortal spirit leaves its domicile of clay its powers of preception, of reason, of reception or rejection of truth or error, of submission or rebellion to the decrees of heaven, are buried with the decaying body? The idea is contrary to all the hopes of the life to come kindled in the heart by the promises of the gospel. It is also antiscriptural. There is nothing in holy writ which establishes any such absurdity. Paul declares that all men shall be judged by the gospel which he preached. If this is true and God is just, must not all men *hear* that gospel and have the opportunity of receiving or rejecting it? And if this privilege has not been granted to them while in the body, must it not be afforded them when out of the body?

Peter states that the Lord shall "judge the quick and the dead," and explains that "For this cause was the gospel preached also to them that are dead, that they might be judged according to men in the flesh, but live according to God in the spirit." He mentions this in connection with his history of the mission and works of Jesus, who, he tells us, was "put to death in the flesh, but quickened by the spirit: by which also He went and preached unto the spirits in prison."

This accounts for the whereabouts of the Savior during the interval between his death on the cross and His resurrection from the sepulchre in the rock. At His appearance to Mary in the garden, after He had risen, He said, "I am not yet ascended to my father." During the three days of His body's sleep in the tomb He was continuing the work the Father had given Him to do. He was preaching "deliverance to the captives, and the opening of the prison to them that were bound."

That these spirits in prison had been in the flesh, Peter makes clear by stating that they were "disobedient * * in the days of Noah,

40

while the ark was a preparing." The gospel was thus preached also to the dead, that they might have the same opportunities and be judged by the same gospel as the living.

The exercise of faith is an operation of the spirit of man, and so is repentance. These lead to obedience and obedience to acceptance with God. The body without the spirit is dead and can neither believe, repent nor obey, but the spirit without the body is active, sentient and capable of exercising all of its powers that are adapted to a spiritual sphere. It is only through the medium of the body, however, that the spirit can handle, experience and fully control or be subjected to corporeal things. That part of the gospel which pertains to earthly ordinances and observances is, therefore, unapproachable to the disembodied. But they can learn and submit to all its spiritual laws and influences and "live according to God in the spirit." They can hear the gospel, for Christ preached it to many of them; they can obey, for He not only proclaimed liberty to them but "He led captivity captive," and they must therefore have repented and become acceptable to God. As one of the early fathers of the Church said of the slain Redeemer, "He went into hades alone, but he came forth with a multitude."

The Jews of Christ's day believed that there were two divisions of the spirit world—Paradise and Tartarus. The good went to the former, the bad to the latter. Jesus promised the repentant thief on the cross: "To-day shalt thou be with me in Paradise." This is not the abode of the Eternal Father but of departed spirits, where they wait until the resurrection. A place of instruction and preparation, of peace and rest, of joy and serenity, of progress toward perfection. And into this abode of the just, Christ led from Tartarus the spirits purified and chastened through their captivity, who were disobedient in the flesh in the days of Noah, but had suffered for their rebellion, and in the spirit had gladly received the gospel through His ministrations.

And thus, in the due time of the Lord all who have dwelt upon the earth in any age, Jew, Gentile, heathen, Christian, may hear the glad tidings of the everlasting gospel preached by those appointed and authorized, and have an opportunity of repentance, improvement and reconciliation. But the ordinances which belong to the sphere of mortality cannot be received in a spiritual estate; they belong to the flesh and must be attended to in the flesh. Consideration of the means provided by Infinite Goodness through which the benefits of those essential ordinances can be obtained by believing, repentant, disembodied persons, must be left till the unfolding of another leaf.

NINTH LEAF.

Decrees of God Fixed in the Spiritual as in the Natural Universe—Ordinances Essential—The Living may be Baptized for the Dead—The Principle of Proxy—The Place for the Administration of Vicarious Ordinances—Revelation of Elijah, the Prophet—Connection with the Spirit World—True Order of Communication—Blessed Results of Work Done for the Dead.

The divine fiat has gone forth that "Except a man be born of water and of the Spirit, he cannot enter into the kingdom of God." This is a fixed law. The same certainty that is exhibited in the government of the material universe obtains in the spiritual domain, and is as much a necessity in one as in the other. As man cannot change the revolutions of the planets nor alter the principles that underlie all motion and regulate all matter, so he cannot turn aside the decrees of Jehovah, nor modify, in the least degree, any rule or commandment pertaining to the everlasting gospel. Neither will He who reigns in the unseen world, as well as in the sphere perceived by the senses, swerve from His established laws in the former any more than in the latter.

Baptism, or the birth of water in the form and mode already described, is an essential ordinance. There are others equally necessary in their time and place in the divine plan of human redemption. They must be rightly received and administered, or the blessings that spring from them, as their natural fruit, cannot be enjoyed. As aliens cannot be admitted to the rights and privileges of citizenship in an earthly government, without complying with the naturalization laws in such case made and provided, so aliens from the heavenly kingdom cannot be received into its dominion, nor be adopted into the family of the Eternal King, without obeying the laws set as the conditions of admission.

43

These laws and ordinances will be made known to the inhabitants of this planet, either in the flesh or in the disembodied condition. They will have the opportunity of receiving or rejecting them on the agency given to man, that a just judgment may be rendered in the great day of accounts. But ordinances, such as baptism, the laying on of hands for confirmation, ordination, marriage, etc., belong to the corporeal sphere. They are set for the state of probation.

Water is an earthly element, or compound of elements, and the blessings ordained to flow from the death, burial and new birth, typified by authorized baptism therein, cannot be secured in any other way. Millions of earth's sons and daughters have passed out of the body without obeying the law of baptism. Many of them will gladly accept the word and law of the Lord when it is proclaimed to them in the spirit world. But they cannot there attend to ordinances that belong to the sphere which they have left. Can nothing be done in their case? Must they forever be shut out of the kingdom of heaven? Both justice and mercy join in answering "yes" to the first and "no" to the last question. What, then, is the way of their deliverance?

The living may be baptized for the dead. Other essential ordinances may be attended to vicariously. This glorious truth, hid from human knowledge for centuries, has been made known in this greatest of all divine dispensations. It is indeed light in the midst of darkness. It shines in the depths of the shrouded past, illuminates the mystic future, and reveals the infinite love of God and His tender mercy over all His works. It explains the meaning of scripture texts long considered difficult and obscure. It links by loving ties the living with their dead. It shows why the fathers "without us cannot be made perfect." It opens the way of redemption for the hosts of departed heathens. It brings together in one all who are in Christ, even though parted by the vail that is drawn between the physical and spiritual spheres. It gives men and women the power to become "Saviors on Mount Zion," Jesus being the great Captain in the army of redeemers.

In God's house all things are done in order. There is a right way and a proper place for the administration of ordinances for the dead. The living relatives of those who have departed without an opportunity of obeying the earthly requirements of the plan of salvation, if they have themselves been born of the water and of the spirit, stand in the name and place of the departed and receive the ordinances to be placed to the credit of the dead. Either sex represents its own. Men are not baptized for women, nor women for men. The first-born son in each family has rights of priority connected with this vicarious work if he has proven himself worthy. The ordinances must be administered by those having authority, being set apart for the work, and must be duly witnessed and properly recorded. The books on earth must tally with the records in heaven.

The place for these administrations is in a temple built to the Most High God, after the pattern revealed. The baptismal font, like the brazen sea in the temple of Solomon, is placed in the basement, under the place where the living are wont to assemble, typifying the place for the dead, all things spiritual having their correspondence with things natural. That which is done on earth, according to the divine instructions, is acknowledged in heaven, and is of force and effect in the world to come. Herein is manifested the power of the Holy Priesthood, loosing or binding on earth, and it is loosed or bound in heaven, all according to the commandments and revelation of the Most High through Jesus the anointed.

This principle of proxy runs like a thread of gold throughout the entire robe of salvation. Christ is the proxy of blood for the whole race of sinners. The Spotless One died in the place of the impure. He is the offering for the deadly sin of Adam. He is the propitiation for the evil deeds of a world. The lamb on the smoking altar, the scapegoat turned into the wilderness, the sprinkling of atonement, all the sacrifices of the old covenant, as well as the infinite one of the new, are based on the doctrine of vicarious action and the divine acceptance of authorized substitutes.

The manifestation of this truth in the last dispensation came from the Prophet Elijah in the temple built to the Almighty by the Latter-day Saints in Kirtland, Ohio. On the third of April, 1836, he who was caught up to heaven without death, appeared to Joseph Smith and Oliver Cowdery, and committed the keys of the power to "turn the heart of the fathers to the children, and the heart of the children to their fathers," that the earth might be saved from a curse. The living are thus authorized, under prescribed conditions, to act for the dead, and the fathers in the spirit world look to the children in the flesh to perform for them the works which they were unable to attend to while in the body.

Here is the peculiar blessing upon the heads of the Saints in the grand, culminating and completing dispensation of the fullness of times. To labor for the redemption of their progenitors until every lost link in the line of their ancestry, back to the Abrahamic stock from which they originally sprang, shall be taken up and welded into the perfect family chain. Herein is seen one of the blessings attending the perpetuation of a man's name in the earth; to die leaving no seed being considered in olden times, among the people of God, one of the greatest of calamities. Indeed the glory and dominion, and joy and rapture of the future state will be found to have intimate relation to the family condition, and the promise to Abraham of a numerous posterity was not merely of earthly portent, but reached into the exaltation and beatitudes of eternal existence.

This glorious doctrine bears the key to the sphere within the vail. It regulates the communion of the living with the dead. It saves those who receive it from improper and deceptive spirit communications. Tidings to the living from their friends who have passed away do not come in disorder and confusion, nor by the will of men or women, whether corrupt or pure. Order is maintained in all the works and ways of God. Knowledge that is needful concerning the spiritual sphere will come through an appointed channel and in the appointed place. The temple where the ordinances can be administered for the dead, is the place to hear from the dead. The Priesthood in the flesh, when it is necessary, will receive communications from the Priesthood behind the vail. Most holy conversations on all things pertaining to the redemption of the race, belong in the places prepared in the temples.

The Saints in the flesh are required to use all due diligence in obtaining their genealogies by the means at command, and a spirit has moved upon men in the world to collect and perfect and publish the records of their ancestors, by which, thousands upon thousands of acceptable names have been obtained, and the work of vicarious baptism already done is immense. But that which remains to be accomplished is so vast, that no mind, unless illuminated by the light of God, can see how it can ever be performed and perfected. Yet it will be done, and blessed are they who aid in the heavenly labor! With what joy will they be greeted by the spirits of their progenitors when they meet them in Paradise! What honor will crown their brows in the day of reward and compensation! They will stand among the saviors, and shine among their kindred who are redeemed, like glorious suns in the heavenly constellations!

This divine plan of vicarious action, is one of the broadest, brightest and loveliest leaves in the blessed tree of life. It bears a healing balm for millions upon millions of earth's sons and daughters who have passed away without hearing the only name whereby man can be saved, or who, having heard, were never taught the way of salvation as ordained through Jesus Christ. It is redolent of the love and mercy of the Eternal Father, and bears the sweet perfume of charity and gratitude of the children reaching out after the fathers, of the fathers blest in the works of the children, and of kindred affection enlarged, cemented and perpetuated for ever and ever. It parts the vail between the physical and the spiritual, it softens the heart, and brings the living and the dead nearer to God, and it sanctifies the soul to obedience, worship and devotion, filling it with reverence and adoration of Him who has devised this broad and universal plan for the redemption of the human race.

TENTH LEAF.

Universality of Death—Results of the
Transgression of Law—Dissolution of
the Body not the End of Existence—
What is Resurrection?—The Spiritual
Body of Jesus—All to be Raised from
the Dead—The Order of the
Resurrection—Necessity of an Immortal
Body—Ignorance of the Laws of Nature
—Matter Indestructible—Possibilities of
Creative Energy—Life and Immortality
Brought to Light.

Death is the common heritage. It is a legacy to all the children, left
by our first progenitor. It is the result of transgression, the penalty
of violated law. The immortal pair who dwelt in Eden fell into
mortality through sin. Immortality is the power of continued
existence. But "all things are governed by law." Sin is law-breaking.
To live for ever requires perpetual obedience to the laws of
everlasting life. "That which is governed by law is preserved by
law." By the same rule reversed, the reverse obtains. Therefore,
that which is immortal and obeys not the laws of immortality, will
become mortal. If obedience insures preservation, disobedience
involves destruction. Law reigns in the highest as well as in the
lower spheres of being. Eternal life involves eternal compliance
with the laws of existence.

All seeds produce their own kind. Mortal beings beget mortality.
When the parents of our race became mortal through breaking the
law of their immortal condition, they brought death to their
offspring as well as to themselves. "In Adam all die." The curse of
death smites the whole family. "It is appointed unto man once to
die." No ingenuity he can exercise or precautions he can adopt will
avert the impending doom. The decree has been proclaimed, "Thou
shalt surely die," and it is irrevocable. The taint that came from the
tree of death whose fruit was forbidden, descends to all

generations, and every variety of form and feature, and color and stature, and tendency and peculiarity, have the one common characteristic, the certainty of death.

But is the dissolution of the body the end of existence? Not at all. We have seen that the part of man that comes from heaven lives on when that which comes from the earth returns to the earth. Yet this is not sufficient. The query arises, Shall this body, made mortal through transgression, remain for ever under the penalty of the broken law, or are there some means of expiation for the sin, and restoration from the doom, its consequence? Are all the associations formed in the flesh and pertaining to this mortal state, to perish with the decayed body and be scattered like the dust to which it is resolved? Are the fond relations of husband and wife, and parent and child to be dissolved forever? Is this exquisitely, "fearfully and wonderfully" formed mechanism, with the experiences of its temporal existence, to be obliterated and lose its identity in the material universe?

The answer comes down from the remotest ages, like sweet and sacred music whose tones swell and increase as the chorus is joined by the voices of the prophets and saints of each succeeding dispensation, until the grand harmony thrills every respondent soul. The burden of the song is in the words of the poetic Isaiah: "Thy dead men shall live, together with my dead body shall they arise. Awake and sing, ye that dwell in dust: for thy dew is as the dew of herbs, and the earth shall cast out the dead." And the ringing tones of Job the ancient are heard as a solo whose melody reaches unto heaven: "I know that my Redeemer liveth, and that he shall stand at the latter day upon the earth; and though after my skin worms destroy this body, yet in my flesh shall I see God!"

The faith of all people who have communed with God or have been inspired by the Holy Ghost, has been that they should be resurrected from the dead. They not only had the assurance of spirit life beyond the grave, but of the revivification of the material body. The signification of the word "resurrect" is "to stand up again." That which was laid down was to be raised up. The release of the immortal spirit from the mortal body would not answer to this. It was this mortal that was to put on immortality, this

49

corruptible that was to put on incorruption.

To make this matter certain, Jesus, who expiated the primal sin, after being offered on the cross as the great sacrifice, gave up the ghost. His lifeless body was taken down, embalmed and buried in a new tomb hewed out of the rock. It was guarded by Roman soldiers. On the third day from the interment that body came forth alive from the grave. The same Jesus who was crucified appeared again among His disciples, and proved that the same body interred was brought forth again, by exhibiting the wounds made by the nails and the spear, by permitting them to touch Him, by eating and conversing with them, and by repeated visits.

This was not a mere manifestation of the immortality of the soul, but a demonstration of the resurrection of the body. Yet that body was transformed. The corruptible blood was purged from the veins, and incorruptible spiritual fluid occupied its place. It was buried a natural body, it was resurrected a spiritual body. Here then, was a pattern of that which is to come. This was the "first fruits of them that slept," a glorious sample of the great harvest of the summer of redemption.

Now the sacrifice of the Savior had as one of its chief objects the restoration of mankind to the condition lost by the fall. "As in Adam all die, even so in Christ shall all be made alive." Death came to the race through one man's sin; life comes to the race through one man's atonement for that sin. The remedy is as broad as the disease. The plan is perfect. This is why Christ is called "The resurrection and the life." By virtue of His triumph over sin and His voluntary submission to death, which had no valid claim upon Him, being sinless, He obtained the keys of redemption for all the sleeping dust of the Adamic family. So He made no idle boast or mystic figure of speech when he declared, "The hour is coming, in which all that are in the graves shall hear His voice, and shall come forth; they that have done good, unto the resurrection of life; and they that have done evil, unto the resurrection of damnation."

The raising of the dead, though universal, is not simultaneous. When Christ, who is our life, shall appear, He will first redeem those that are in Him. Having put on Christ and received of His spirit, they will come forth at His call to meet Him. They who have part in the first resurrection are those who have died in the Lord and are blessed and holy. Their bodies will be fashioned like unto His glorious body. Having been planted in the likeness of His death they will be also in the likeness of His resurrection. That is, they will be quickened by the celestial glory and be placed in a condition to receive a fullness thereof, and inherit all things as joint heirs with Christ.

The wicked dead remain unquickened for a thousand years. They reap the fruits of their evil seeds sown in lives of transgression. They drink the dregs of a bitter cup. Some are beaten with many stripes, others with but a few. Justice metes out to them their dues. And when they come forth to stand up in their bodies, they will not be quickened by the celestial glory, but by that for which they are fitted by their respective conditions consequent upon their earthly acts, and they will occupy positions accordingly. But all will be redeemed in due season from the grave and stand the scrutiny of the All-Seeing Eye and the judgment of unswerving Justice, which will determine their eternal future.

In this age of general doubt, when human reason is exalted above divine testimony, and the voice of faith is drowned by the clamors of pretended science, the possibility and use of a resuscitation of the body are scouted and denied. But "all things are possible to them that believe," and the divinely illuminated mind can perceive not only the use, but the necessity of the resurrection.

The being that was placed in Eden and endowed with power to wield dominion over all created things, was a living soul, a sentient spirit in an immortal body, a man fashioned in the image of God. He fell from that condition and paid the penalty of death. Christ's atonement, as we have seen restores him to his original condition. But this he cannot have without his body again made immortal. By the workings of the grand scheme of human exaltation, he and his posterity, with the benefits of the lessons of experience, will be restored to the immortality and pleasures of the primeval paradise, and placed on the path of eternal progress.

And, mark this, a body framed out of the grosser elements is

51

essential to the perfect happiness and power of the refined spiritual organism which possesses it as a tabernacle. The principle of affinities and of the attraction and communion of similars proclaim this truth. Spirit ministers to spirit. Things of a like nature cohere. The higher or spiritual element reaches upward to the loftiest things; the lower or fleshy element reaches downward, and the twain, inseparably combined and governed by the laws of right and truth, draw pleasure and delight from the heights and depth of the boundless universe and the ever-extending spheres of eternal intelligence. A disembodied spirit is imperfect, and requires clothing with its denser parts. Without them, its affinities would lie in but one direction, and its joy and progress would be limited.

The family condition too is formed in the embodied state. Death separates the husband and wife, the parents and children. The resurrection, in its highest conditions, reunites them and restores all that was lost in the grave. Who can picture the bliss, the glory, the power, the might, the dominion and majesty that shall grow out of the redemption from the dead of the righteous man and his household, dwelling in perfect harmony and peace with all the powers of their being, spiritual and physical, purified, quickened, intensified and enlarged to a fullness, with all eternity before them for the exercise thereof in accordance with the designs of the Great Greater? It is beyond the skill of man to depict it, and no mortal mind can comprehend it without special divine illumination.

And who shall define the impossible, or draw the bounds of the powers of the Creator? The secret of ordinary life is hidden from the scrutiny of the most profound scientist. He knows not the mystery of the vital principle that quickens even the lowest form of animated nature. His own powers of mind and motion are incomprehensible to him. Their origin and cause are beyond his ken, and he cannot solve the problem any better than the ignorant Hottentot or the untutored Indian. The reproduction of plants from their seeds, the evolving of life out of the midst of their death, is a wonder unexplained. And shall we say that it is impossible for the Power that regulates the universe to reanimate a defunct body?

It must be remembered that nothing in nature is annihilated. No particle of matter is destroyed by any process. What is called death is but a change of form. All matter is not visible to the human eye. A body may exist, but so transformed as to be imperceptible to the natural vision. The forces that regulate the universe are occult, and though some of the laws that govern them are known, there are others that have not been discovered, and it is the height of presumption for those who have obtained a smattering of information concerning these things—and who has obtained more? —to declare that impossible which they know nothing of, or to limit the power of that creative or quickening energy, whose nature, capabilities and qualities they cannot comprehend in the smallest degree.

If one dead body has been raised to life, unnumbered millions may also be revived. That one we have in the person of Jesus of Nazareth, and He is the forerunner of all the race. Let the sons and daughters of men rejoice and give thanks to Him who has wrought out this great redemption. Death is conquered. The grave has no terrors. Life and immortality are brought to light. Eternity with all its prospects and capabilities is open to the view. And through the power of the resurrection vested in Christ Jesus, the whole globe shall deliver up its dead, and the great progenitor of our race, Adam, the "Ancient of Days," shall stand forth at the head of his posterity all quickened and animated by the spirit of life; and while Jesus the Son is hailed as the mighty Redeemer, God the Eternal Father shall be honored and worshiped for ever as the Author of our being, from whom springs all life, light, power and glory throughout the vast domains of universal space!

ELEVENTH LEAF.

Man or Woman Alone Imperfect—
Marriage Ordained of God—Sanctity of
Proper Sexual Relations—Matrimony a
Part of Religion—The First Pair
Immortal—Marriage for Eternity—Keys
of Celestial Marriage—Condition of
Those who Marry Only for Time—Man
the Head of the Woman—Plurality of
Wires—Continuation of the Righteous
Forever—Eternal Family Organizations
—Everlasting Increase and Dominion.

No man or woman, separate and single, can attain the fullness of
celestial glory. Perfection of being, happiness, exaltation or
dominion, is unattainable by either sex alone. The nature, desires,
capabilities and manifest design of both male and female humanity
proclaim this, and the voice of Deity has endorsed and sanctified
the utterance of nature. Woman was made for man. Marriage is
ordained of God. In its correct form it is under the divine direction.
The Father of the race has the right to a voice in the sexual unions
of His children. Those relations are fraught with so much
consequence, relating to time and eternity, that the Supreme Ruler
should regulate them for the benefit of the parties, the welfare of
society and the good of posterity in this world, as well as for
eternal results in the life to come.

The male and female elements of humanity seek union, of their own
volition. The natural attraction that prompts this is right and proper.
But if there were no rules and restrictions for the government of
these tendencies and the actions resultant, confusion would ensue,
and the effects would be sorrow, ruin and destruction. Matrimony
therefore becomes a part of religion. It is a divine institution, and
hence should be divinely directed. The first marriage on record was
solemnized by Deity. It was God who said, "It is not good that the
man should be alone." It was God who brought Eve and gave her

to Adam. It was God who commanded the twain made one flesh to "increase and multiply."

Marriage, properly contracted, is therefore holy and pure, and its relations, unabused, are sacred and chaste. The notion that celibacy is purer than matrimony, that either man or woman is holier in the sight of heaven because of nonintercourse with the other sex, is a gross error, unwarranted by reason or revelation. There is no attribute of the mind or function of the body that is in itself, or in its legitimate exercise, impure or degrading. It is only the wrong use of any of our powers that is sinful.

The first marriage recorded in scripture was the union of immortals. The curse of death had not been pronounced when the ceremony was solemnized. There was no sin then, and therefore there was no death. The man and woman became ONE as eternal beings, and dominion was given to them over all earthly things, together. Death and the rule of man over the woman came as the consequences of transgression. The penalty was paid, the redemption was wrought out, and through the atonement those two persons are restored to their pristine condition.

In the resurrection, then, Adam and Eve come together as at the first in the garden, and there is no more separation for them. They are rejoined, not as ghostly beings without the feelings and powers of tangible personality, but as the man and the woman made one eternally, with power to increase and multiply and have dominion, with all eternity before them for the exercise of every power with which the Creator endowed them, spiritual, mental and physical, standing at the head of the race, perfected by experience and obedience to eternal law, and ready to act in the harmony with celestial intelligencies, and preside over their own posterity forever.

Here is a sample marriage. It was not for time alone, but for eternity. Death intervened, but only as an incident. The bond that bound them in matrimony was not sundered. The seal set upon them was of heavenly stamp. Its virtue reached within the vail. Its force extended into the world to come. There was no end to it. God had a hand in it and it was His seal and sanction that made it valid and everlasting. All other marriages solemnized on similar

55

principles and under the same authority will be of the same virtue and effect. Ordinances performed by those divinely appointed are as though attended to by Deity in person. "Whoso receiveth you receiveth me," saith the Lord. What they "bind on earth shall be bound in heaven." Herein is the authority of the holy Priesthood, and herein is the sealing power for the Saints of God, by which they may enter into the holy order of celestial marriage that lasts while eternity endures. The KEYS of this power are only held by one man at a time on the earth, being vested in the president of the whole Church of God in the flesh. But while he holds the keys, others may officiate therein under his direction and authority.

Unions formed by men and women, of their own arrangement without any divine sanction or divine ceremony, are only temporary in their nature. They end when the parties or either of them die. God does not acknowledge that which He has not appointed. Neither the vows of the man and woman, nor the ceremony performed by a person unauthorized by the Almighty are recognized in heaven, but only pertain to earth and time. The claim of parents thus united, over their offspring, is but of the earth, earthy, and does not extend into the spheres beyond. Death dissolves both these marital and parental ties, and each family particle becomes disintegrated.

No power but that of Deity can bring them again together, and as God proceeds by law, and the law fixed for these relations have not been complied with, the separation continues while endless ages roll. "In the resurrection *they* neither marry, nor are given in marriage," but, if in a saved condition, are as the angels, and they are ministering spirits or servants unto those who obtain the crown of eternal lives, "a far more and exceeding and eternal weight of glory," than that which rests upon any of the angels. Men and women may be *saved* in a separate and single state, but they cannot be *exalted* into the fullness of celestial glory without union in celestial marriage, because that is a state of perfection and comprehends the gift of perpetual increase, in which there are endless dominion and the exercise of all the powers of immortal manhood and womanhood united as one in the everlasting covenant.

In the divine economy, as in nature, the man "is the head of the woman," and it is written that "he is the savior of the body." But "the man is not without the woman" any more than the woman is without the man, in the Lord. Adam was first formed, then Eve. In the resurrection they stand side by side and hold dominion together. Every man who overcomes all things and is thereby entitled to inherit all things, receives power to bring up his wife to join him in the possession and enjoyment thereof.

In the case of a man marrying a wife in the everlasting covenant who dies while he continues in the flesh and marries another by the same divine law, each wife will come forth in her order and enter with him into his glory. Is there any reason why this should not be so? Is not each of these wives entitled to her position in eternity, by virtue of the sealing power which made her part of the man? Why should one enter into the exaltation of the celestial world, and the other be relegated to singleness and servitude? They all become one in the patriarchal order of family government. And if this be the case in heaven, why should not similar conditions so far as possible exist on earth? Is earth holier than heaven? If a man receives from the Lord more wives than one under the sealing ordinances of celestial marriage, where is the moral wrong? They belong to no other man, but are his by mutual consent of all the interested parties, and they live together in the marriage state, one as much as the other.

In this position there are occasions for the exercise of patience, forbearance, charity, self-sacrifice and the exercise of all the virtues to a far greater degree than in any other. In this plural family relation, an experience can be gained that no other condition in life affords, and the parties who so live and keep the law will be, in the very nature of things, prepared for a wider sphere of dominion, and power, and dignity, and might in the eternal world, than those who have only experienced the monogamic condition. They will, therefore, if they endure unto the end, go forward into the highest degree of exaltation, while their posterity will multiply in an ever-increasing ratio, until worlds will be filled by their generations and they will ascend to the majesty and splendor of the Gods on high.

Herein is our Eternal Father glorified and His dominions extended.

By the continuation of the seeds of the righteous forever, the multiplication of His sons and daughters creates the needs for worlds and systems, to be brought forth according to eternal laws, to occupy their position in the universe as dwelling places for spirits, and embodied mortals, and perfected souls, in the various grades on the path of progress towards the perfection of the celestial order; as orbs of light and splendor, or globes of trial, punishment or correction, each in its allotted sphere in the galaxy of suns and stars and planets, and in the vast and wondrous plans of the Mighty Architect, the Eternal Parent of organized intelligencies.

In obedience to His laws, there is present peace and future joy. They who are in harmony with Him are in affinity with the source of all pleasure and power. His commandments are found in the laws of continuing life, which regulate as permanent things; and they who reject Him and His counsels shut the gate against their own happiness and advancement. But, for them who receive His gospel and conform to all its ordinances and teachings, the door is open to the highest courts in the heavenly mansions, and while they are helped through the ordeals of mortal life, they gain the keys to all the glories of that existence in which the family relation is perfected and perpetuated, and every power of the whole being, refined, intensified and developed, finds exercise, in its true sphere, to the complete and unalloyed bliss of each one in the endless family circle, and the glory of Him who is the Patriarch and Ruler of all.

TWELFTH LEAF.

Christ's Work Continued After His
Death—The Perfect Science of Human
Redemption—What was Lost in the Fall
—What is to be Regained in the
Restoration—Justice Tempered with
Mercy—Loss Sustained by the
Disobedient—Doom of the Sons of
Perdition—The Celestial, Terrestrial and
Telestial Glories—Redemption and
Glorification of the Earth—Salvation of
the Whole Race—The Finished Work of
Christ—Universal Dominion of the
Father.

The mission of Christ was to save that which was lost. It was not
completed when He hung upon the cross. His dying exclamation,
"It is finished!" referred to His sufferings for sin, the ordeals of
mortality, His labors in the flesh. As we have seen, He continued
His work of salvation when out of the body, by preaching to the
dead. After His resurrection He met, on several occasions, with His
disciples, and instructed them in the plan of redemption and sent
them to all nations, that the work He had commenced on earth
might be continued. He ministered to other nations, uttered His
voice to other sheep which were not of the fold in Palestine, that
the lost tribes of Israel and all who could not be reached by His
Jewish Apostles might hear the glad tidings of salvation. This,
however, not fully revealed in the Bible, is made clear in the Book
of Mormon. After His ascension, to fulfill His own promise, He
went to prepare a place for His faithful disciples, that when they
left the earth they might be able to abide with Him.

But all this was only a small part of the perfect scheme of
redemption. That which was lost in Adam is to be regained in
Christ. Through the commission of crime, death came into the
world. Satan gained dominion. The earth trembled under the curse.

59

Eden bloomed no more upon its face. The tree of life was removed. Thorns and briers and noxious weeds came up in the place of the flowers and fruits of paradise. Deity was hidden from the sight of man. Sorrow and pain and toil and travail became the heritage of mortals. Enmity arose between man and beast. Venom entered the serpent's fangs, and rage the hearts of brute and fowl and aqueous creature. Strife dwelt in the very elements and death brooded over the face of the smitten globe.

What, then, was lost? The immortality of man; the blessed tree of life; communion with Jehovah; the companionship of angels; the purity of paradise; man's dominion over inferior creatures; freedom from satanic influence; exemption from toil and pain; earth's affinity with perfected realms on high.

Until all this has been restored, Christ's work must continue. The earth must be cleansed from its corruptions. The elements must melt with fervent heat, and be purified from evil. Satan and his hosts must be banished and bound. Eden must blossom again as at first. The lion and the lamb must lie down together. The fig tree and the myrtle must flourish where the rank weeds grow. The whole race of Adam must be raised from the dead. The vail between earth and heaven must be removed. The knowledge and glory of God must cover the earth as the waters cover the deep, and the spirit of life and peace and light and joy must be poured out upon all flesh, until the whole creation vibrates with pleasure and responds with praise.

The ushering in of the great millennial day, a glimpse of which has been seen by all the holy prophets since the world began, with the sweet rest of earth and its inhabitants, is not, however, the completion of Christ's glorious work. His kingdom must not only be established from pole to pole and from shore to shore, but His saving power must penetrate to every lost soul of our race in, the regions of the damned.

A just judgment will be meted out to all. They who reject the gospel must suffer the penalty. Those who are found worthy of many stripes must receive their portion. The wicked will be turned into hell, with all the nations that forget God. Each condemned person will pay the uttermost farthing for his sins. Justice, tempered, not warped or thwarted, by mercy, will mete out to all their right deserts, "every man according to His works." The punishment is always existent, therefore it is eternal punishment. But each one who suffers, receives only his just portion thereof. Shall the murderer and the Sabbath-breaker, the adulterer and the thief, the drunkard and the profane, all merit the same doom? Would human courts proclaim such judgment? Shall man have more equity than God? When stern justice has claimed its own and filled its purpose, shall there be no place for sweet mercy?

While there is one soul of this race, willing and able to accept and obey the laws of redemption, no matter where or in what condition it may be found, Christ's work will be incomplete until that being is brought up from death and hell, and placed in a position of progress, upward and onward, in such glory as is possible for its enjoyment and the service of the great God.

The punishment inflicted will be adequate to the wrongs performed. In one sense the sinner will always suffer its effects. When the debt is paid and justice is satisfied; when obedience is learned through the lessons of sad experience; when the grateful and subdued soul comes forth from the everlasting punishment, thoroughly willing to comply with the laws once rejected; there will be an abiding sense of loss. The fullness of celestial glory in the presence and society of God and the Lamb are beyond the reach of that saved but not perfected soul, forever. The power of increase, wherein is dominion and exaltation, and crowns of immeasurable glory, is not for the class of beings who have been thrust down to hell and endured the wrath of God for the period allotted by eternal judgment.

But Jesus, the anointed, with His army of saviors bearing the Priesthood after the order of Melchisedec, will seek and save that which is lost until everything salvable is redeemed. Only those beings who have learned the law, received of the light of truth, tasted the sweets of the divine spirit, basked in the sunbeams of the heavenly glory, made covenant to serve the King of kings and received power to advance to the pinnacle, of exaltation, and then have turned away from the right, chosen evil rather than good, driven away the power and promptings of the Spirit of light and truth, sought to become a law unto themselves, imbrued their hands in the blood of innocence or, drinking in of the influence of that evil one, consented to and endorsed the slaying of the world's Redeemer, thus sinning against the Holy Ghost and becoming servants of Satan and sons of perdition, will be in their nature and status unredeemable, and therefore will remain "filthy still" and thus be unfit for a kingdom of any degree of glory. Those will go away with the devil and his angels into the outer darkness, beyond the spheres where flows the river of salvation and where blooms the tree of life. For them alone of Adam's race there is no repentance, for them alone is the second death, for them alone is the blackness of darkness forever.

When the work of Christ and His associate kings and priests unto God is finished, the saints of all the ages will be crowned with glory and receive their reward. They will be made rulers over many things. In the order of eternity, they will stand in the heavenly family organization, and all things will be theirs. Of their increase there will be no end. They will hold the key to all heights and depths. They will have power over all the elements, spiritual and corporeal. The incorruptible and fadeless riches will be theirs. They will mingle with the highest. They will gaze upon the face of the Eternal God and dwell in the presence of the sinless Son. Pain and sorrow, and trial and death will henceforth be only known in memory, to form the contrast needful to make their joy complete. Eternity with its boundless opportunities and unutterable bliss and intelligence and majesty will be before them without a barrier in the way, secure for them as to the Almighty Father himself. This is the celestial glory.

Those who were not numbered with the Saints of God in the flesh, but who received the gospel in the spirit; the good and honorable who were led astray by the designing; the class not fitted for the crowning glory of the celestial world nor worthy of the doom of the wicked, will also receive their portion. They will not attain to the gifts of increase and dominion and the fullness of the highest, but will enter into their rest, which shall be glorious. And though they reach not to the Father's fullness, they will receive the visits of the Son and of His associates in the celestial world, and enjoy rich blessings unspeakable in their greatness and perpetuity. They inherit the terrestrial glory.

Those who were cast down to the depths for their sins, who rejected the gospel of Jesus, who persecuted the saints, who reveled in iniquity, who committed all manner of transgressions except the unpardonable crime, will also come forth in the Lord's time, through the blood of the Lamb and the ministry of His disciples and their own repentance and willing acceptance of divine law, and enter into the various degrees of glory and power and progress and light, according to their different capacities and adaptabilities. They cannot go up into the society of the Father nor receive of the presence of the Son, but will have ministrations of messengers from the terrestrial world, and have joy beyond all expectation and the conception of uninspired mortal minds. They will all bow the knee to Christ and serve God the Father, and have an eternity of usefulness and happiness in harmony with the higher powers. They receive the telestial glory.

Thus the inhabitants of earth, with the few exceptions that are beyond the power of redemption, will eventually be saved. And the globe on which they passed their probation, having kept the law of its being, will come into remembrance before its Maker. It will die like its products. But it will be quickened again and resurrected in the celestial glory. It has been born of the water, it will also be born of the Spirit. Purified by fire from all the corruptions that once defiled it, developed into its perfections as one of the family of worlds fitted for the Creator's presence, all its latent light awakened into scintillating action, it will move up into its place among the orbs governed by celestial time, and shining "like a sea of glass mingled with fire," every tint and color of the heavenly bow radiating from its surface, the ransomed of the Lord will dwell upon it; the highest beings of the ancient orbs will visit it; the garden of God will again adorn it; the heavenly government will prevail in every part; Jesus will reign as its King; the river of life will flow out from the regal throne; the tree of life, whose leaves were for the healing of the nations, will flourish upon the banks of the heavenly stream, and its golden fruit will be free for the white-robed throng, that they may eat and live forever. This perfected earth and its saved inhabitants will then be presented to the Eternal Father as the finished work of Christ, and all things will be subject unto the Great Patriarch, Architect, Creator, Ruler, the Almighty, to whom be obedience and reverence and praise in all the countless worlds that shine as jewels in His universal crown!

THE LATTER-DAY KINGDOM.

How shall I sing thy beauty, pow'r and light,
  O glorious kingdom of the latter days!
I see thy loveliness, I feel thy might,
  But find no utterance to speak thy praise!

I search in vain the records of the past,
  Which paint dead kingdoms in their short-lived pride,
They cannot picture thee, whose pow'r shall last
  While heav'n and truth and Deity abide.

And shall the little "powers that be" to-day,
  Be likened for a moment to thy majesty?
As well declare pale vesta's twinkling ray
  Unfolds the splendor of eternity.

In hist'ry only Egypt's greatness lives—
  Lost are its treasures, all its wisdom hid,
Except the scraps the crumbling mummy gives,
  The sculptured sphynx and tow'ring pyramid.

Assyria! Thy sceptre lies in dust.
  Thy bow is broken and thy pomp has fled.
Perished thy fruits of conquest, blood and lust,
  With all the warriors that Tiglath led!

Where are the palaces of Babylon,
  The "hanging gardens" and the golden tow'rs?
With the Chaldeans' starlight wisdom gone,
  Walls, gates and glory, images and flow'rs!

And couldst not thou, Greece, avert thy fate,
  With oracles and wealth and victory?
Couldst not thy world-wide reign perpetuate,
  With all thy gods and deep philosophy?

The soul that moved thee in thy conquering march,

That spoke in poesy and art and grace,
Is disembodied; and the mouldering arch
  And chiseled fragment mark thy burial place.

And thou, O Rome! proud mistress of the world!
  Thine armored legions spread no terror now.
They bring no blood-bought spoils of gems impearled,
  To deck thy bosom and thy haughty brow.

Thy Coliseum's vast and vacant walls,
  Rot as an emblem of thy great decay,
And on the ear its mournful echo falls,
  A dismal knell of thy departed sway!

O! all ye living governments and states!
  Gaze on the relics of far mightier powers!
The hand that shattered them, uplifted waits
  The bell that ends your few remaining hours!

In the high chambers of the West, I see
  An infant kingdom struggling to the birth.
And the prophetic spirit says to me,
  "In manhood this shall govern all the earth."

O Zion! built by Saints of latter days.
  Bring forth the promised kingdom to the world!
Upon the mountain tops "the ensign" raise,
  And spread its shining folds to all the world!

Gathered from ev'ry clime and tongue and race,
  Under the banner, righteous men shall stand,
And the all-conquering Christ shall show His face,
  And give dominion to that faithful band.

Armored in truth and God's authority,
  Dauntless and terrible, yet full of love,
The King shall lead them unto victory,
  And bring a van-guard from the ranks above.

No weapon formed against them shall prevail,

No cunning plan shall prove their overthrow;
The prince of all earth's kingdoms they assail,
  And drive his forces to the shades below.

The spirit that gives wisdom to the wise,
  Prom Council, Congress, Parliament, shall flee—
Shall rest on those whom all mankind despise,
  And leave the world to human policy.

Left, in a day of storms, each bark of state,
  Rotten and rudderless, whirled madly on
Against each other on the sea of fate,
  With awful crash to depths of death go down.

But see the ship no storm can overwhelm,
  Saving the remnants of the wrecks below!
"The Priesthood" 's written on her shining helm,
  "God's Kingdom" is inscribed upon her bow.

God's Kingdom! seen in vision by the seers!
  God's Kingdom! Clothed in justice truth and light!
Theme of the prophet and the bard appears,
  To save the nations from chaotic night.

A perfect government for all the earth.
  Not a republic nor a monarchy,
And yet from both all principles of worth
  Are blended in this great Theocracy.

Wielding almighty power in ev'ry land,
  The willing people bend to its supreme decrees,
And mutual int'rest, like a golden band,
  Binds in one social compact men of all degrees.

Appointed by the great Jehovah's voice,
  By intellect and virtue qualified,
And a free people's universal choice,
  The leading spirits govern and preside.

No longer bound beneath the cruel weight

Of idle vampires, draining their life's blood,
The joyful nations yield the pow'r of state,
  To legislators for their country's good.

Earth's treasures, hiding 'neath the deep sea waves,
  Bound in the rock, or shining on the strand,
Or glittering in subterraneous caves,
  Come sparkling forth at industry's command.

New sciences and arts diffuse new light,
  Knowledge of future and of past events,
Wisdom to comprehend the secret might,
  And subtle forces of the elements.

In wondrous implements, mechanic skill
  Gives unto labor swift and easy wings,
Making each sterile spot with life to thrill,
  While water from the thirsty desert springs.

Thought, freed from human trammels, brings to light
  Its glorious conceptions without fear,
And mouldy Precedent, struck dead with fright,
  Reposes on an unregretted bier.

The laws which life and health perpetuate,
  By inspiration's sacred voice are taught,
And every passion made subordinate,
  To principles with lasting pleasure fraught.

Jesus, the Sinless, fills the regal throne,
  To Him all other rulers bend the knee;
He reigns not by His right and might alone,
  But loving homage swells His majesty.

Earth linked into the chain of worlds on high,
  Among the ransomed planets takes its place,
And finds itself in blest affinity
  With orbs that govern time through boundless space.

Such is the kingdom now on earth begun,

A branch of the great Governmental Tree,
Whose roots are grounded in the central sun,
  Whose boughs bear fruit through all eternity.

Happy are they who labor in its cause,
  Happy are they who suffer for its sake;
For all who are obedient to its laws,
  Of all its joys and honors shall partake.

APPENDIX

Scripture references in proof of the doctrines set
forth in the body of this work.

FIRST LEAF.

But one God to worship—I. Cor. viii, 6.
Man's ways not accepted of God—Matt. xv, 9.
Only one correct way—John x, 1.
Faith the first principle—Heb. xi, 6.
Faith a principle of power—Heb. x.
How faith comes—Rom. x, 14, 17.
Human learning inadequate—I. Cor. ii, 5-14.
 God the Father of spirits—Heb. xii, 9; Eccles. xii, 7; John
xx, 17.
Man in God's image—Gen. i, 26; I. Cor. xi, 7.

SECOND LEAF.

Death by Adam, life by Christ—Rom. x, 12-21
 All, good and bad, to be raised from the dead—I. Cor. xv,
22;
    John v, 28, 29; Daniel xii, 2.
Christ died for original sin—John i, 29; Rom. v, 18, 19; I.
    Cor. xv. 21, 22.
Christ died for our actual sins—Rom. iv, 25: v, 8; viii, 32;
    I. Cor. xv. 3; Galatians i, 4; Ephesians, i, 7: Collossians,
    i, 14; Heb. ii, 9; ix, 28; I. Peter ii, 24; iii, 18;
    I. John ii, 2.
Faith repentance and baptism fundamental principles—Heb.
    vi, 1, 2; Matthew xxviii, 19, 20;
True repentance—II. Cor. viii, 10, 11.
Baptism is immersion—Rom. vi, 4; Acts viii, 38, 39; Mark i,
4.
But one baptism—Ephesians iv, 5.
But one door into the fold—John x. 1, 2.

THIRD LEAF.

Gift of the Holy Ghost by laying on of hands—Acts viii,
    14-19; xix, 6; II. Timothy i, 6: Deut. xxxiv. 9.
Office of the Holy Ghost—John xiv, 26; xvi, 13.
Fruits of the Spirit-Gal. v, 22, 23.
Birth of the Spirit essential—John iii, 3-5.

FOURTH LEAF.

No man to take the Priesthood upon himself—Heb. v, 1-4.
What is sealed on earth by authority is sealed in heaven—
    Matt. xviii, 18; xvi 19.
The Melchisedek Priesthood eternal—Heb. v, 5.
The Aaronic or Levitical Priesthood another order, Heb. vii,
    11, 21.
Jesus did not assume the Priesthood—Heb. v, 5.
God called Jesus to the Priesthood—Jeb. v, 10; Psalms cx,
4.
Moses and Elias administered to Jesus—Matt. xvii, 1-5;
    Mark ix, 2-7, Luke ix, 30-35.
Jesus ordained apostles to the same authority—John xx,
    21-23; xvii, 22; xv, 16.
The Apostles ordained others—Acts i, 23-26; vi, 6; xiii,
    1-3; xiv, 23; xv, 22; I. Timothy iv, 134; Titus i, 5.

FIFTH LEAF.

Christ the head of the Church—Ephesians v, 23: i, 22.
Apostleship first authority in the Church—I. Cor. xii. 28;
    Ephesians ii, 20.
Peter, James, and John chief Apostles—Galatians ii, 9.
Seventies called and sent forth—x, 1.
Officers of the Church—I. Cor. xii, 28: Eph. iv, 11; I.
    Timothy iii, 1-13; v, 1.
Apostles and Prophets to continue—Eph. iv, 13.
Progress of the Church towards perfection—Eph. iv, 12-16;
    v, 27.
Church casts out evil doers—II Thess. iii, 6-14; Rom. xvi,
    17; I. Cor. v, 4-11; II. Cor vi, 14-17; Matt. xviii, 17.

71

Members of the Church all one; no nationality—I, Cor. xii,
13;

    Galatians iii, 28; Rom. x, 12; Ephesians ii, 19-22.

Church of the present connected with the past—Heb. xii,
    22, 23.

SIXTH LEAF

God not the author of confusion—I. Cor. xiii, 33.

Contention among the early Saints—I. Cor. i, 11.

Great apostasy foretold—II. Thess. ii, 2, 3; I. Tim. iv, 1; II.
    Tim. iii, 1-7; II. Peter ii, 1-3; Revelations xiv, 8.

The iniquity commenced in the first century—II. Thess. ii, 7;
    Rev. ii, 3.

Christ's pure Church symbolized—Rev. xii, 1-5.

The apostate church contrasted—Rev. xvii, 1-6.

Darkness covered the earth—Isaiah lx, 2.

Spirit of the deep sleep poured out—Isaiah xxix, 9, 10.

The world worshipping God only with their mouth—Isaiah
    xxix, 13.

SEVENTH LEAF.

Restoration of the gospel by an angel—Rev. xiv, 6, 7.

Knowledge to follow obedience—John vii, 17.

John the Baptist could not confer the Holy Ghost—Matt. iii,
    11; Acts xix, 2-4.

Powers of the Aaronic Priesthood—Doctrine and Covenants
    Section cvii, 20.

Powers of the Melchisedek Priesthood—Doc. and Cov. Sec.
    cvii, 18, 19; Heb. v, ix.

Signs to follow believers—Mark xvi, 17, 18; I. Cor. xii, 7-11.

Dispensation of the fullness of times—Eph. i, 9, 10.

EIGHTH LEAF.

No salvation but by Jesus Christ—Acts. iv, 12.

Birth of water and of spirit essential—John iii, 5.

All to be judged by the gospel—Rom. ii, 16.

Gospel preached to the dead—I. Peter iv, 6.

Christ preached to the spirits in prison—I. Peter iii, 18-20

Preaching to captives foretold—Isaiah lxi, 1: xlii, 6, 7.

Jesus led captivity captive—Eph. iv, 8.

Jesus did not go to heaven when He died—John xx, 17;
    Luke xxiii, 43.

Living and dead to hear the gospel—Rom. x, 14; Isaiah xxiv,
    21, 22.

NINTH LEAF.

Baptism for the dead—I. Cor. xv, 29.

The fathers without us not perfect—Heb. xi, 39, 40.

Saviors on Mount Zion—Obadiah i, 21.

 Order of baptism for the dead—Doc. and Cov. Sections
cxxvii,
    cxviii.

Elijah the Prophet to come—Malachi iv, 5.

Christ the proxy of blood for all—Heb. ix, 12, 14, 22; x, 10;
    I. Tim. ii, 6.

Knowledge about the dead to some from God—Isaiah viii,
    19, 20.

TENTH LEAF.

Sin the transgression of law—I. John iii, 14.

Death the wages of sin—Rom. vi, 23.

All men to die—Heb. ix, 27; Eccles. iii, 20.

Death inherited from Adam—Rom. v, 12.

Life after death—II. Cor. vi; Eccles. xii, 7.

Resurrection of the body—Job xix, 25-27; Isaiah xxvi, 19;
    Luke xxiv, 26-42; I. Cor. xv, 35-54; Phil. iii, 20, 21.

First resurrection—Rev. xx, 4-6.

Three glories-I. Cor. xv, 15.

A body necessary for full happiness—Ezekiel xxxvii, 2; Doc.
    and Cov. Sec. xciii, 23, 24.

The Ancient of Days—Daniel vii, 9-14.

ELEVENTH LEAF.

Woman made for man—I. Cor. xi, 9.

Marriage ordained of God—Gen. ii, 22-24; i, 28.

Marriage honorable—Heb. xiii, 4.

Man the head of the woman—Eph. v, 23; I. Cor. xi, 3.

Man not without the woman in the Lord—I. Cor. xi, 11.

Unmarried persons as the angels—Matt. xxii, 30.

Saints to judge angels—I. Cor. vi, 3.

Angels to be ministering Spirits—Heb. i, 14.

God gave David wives—Judges viii, 30.

Jacob's four wives—Gen. xxx, 1-26

Abraham and his wives—Gen. xviii, 16-19; xvi, 1-3; xxv,
    1-6

Abraham and Jacob in the kingdom of heaven—Matt. viii,
    11, 12; Luke xiii, 28.

Celestial marriage—Doc. and Cov. Sec. cxxxii.

TWELFTH LEAF.

Christ came to save that which was lost—Matt. xviii, 11.

 Christ ministered to His disciples after His resurrection—
Acts
    i, 3-8; I. Cor. xv, 5-8.

Other sheep besides the fold at Jerusalem—John x, 16.

Christ prepared a place for His disciples—John xiv, 2, 3.

 Earth to be cleansed from corruption—Isaiah xxiv, 1-6;
Malachi
    iv, 1-3; II. Peter iii, 10-12.

Satan to be bound—Rev. xx, 1-3.

Restoration—Isaiah xi, 6-9; lxv, 17-25.

All to be judged according to their works—Rev. xx, 12-15;
    Matt. xvi. 27.

Some beaten with few, some with many stripes—Luke xii,
    47-48.

Pay the uttermost farthing—Matt. v, 26.

The unpardonable sin—Mark iii, 28, 29; i. John v, 16.

The second death—Rev. xxi. 8; xix, 20.

Remain filthy still—Rev. xxii, 11.

The future of mankind—Doc. and Cov. Section lxxvi.

Every knee shall bow—Philippians ii, 10.

Earth to be made new—II. Peter iii, 13; Rev. xxi. 1.

Sea of glass mingled with first—Rev. xv. 2.
The righteous to inherit all things—Rev. xxi, 7.
The river of life—Rev. xxii, 1; Ezekiel xlvii, 1.
Leaves of the the tree of life—Rev. xxii, 2.
All things become subject to God—I. Cor. xv. 24-28.